VEGETABLES

Susan Wake

Illustrations by John Yates

Carolrhoda Books, Inc./Minneapolis

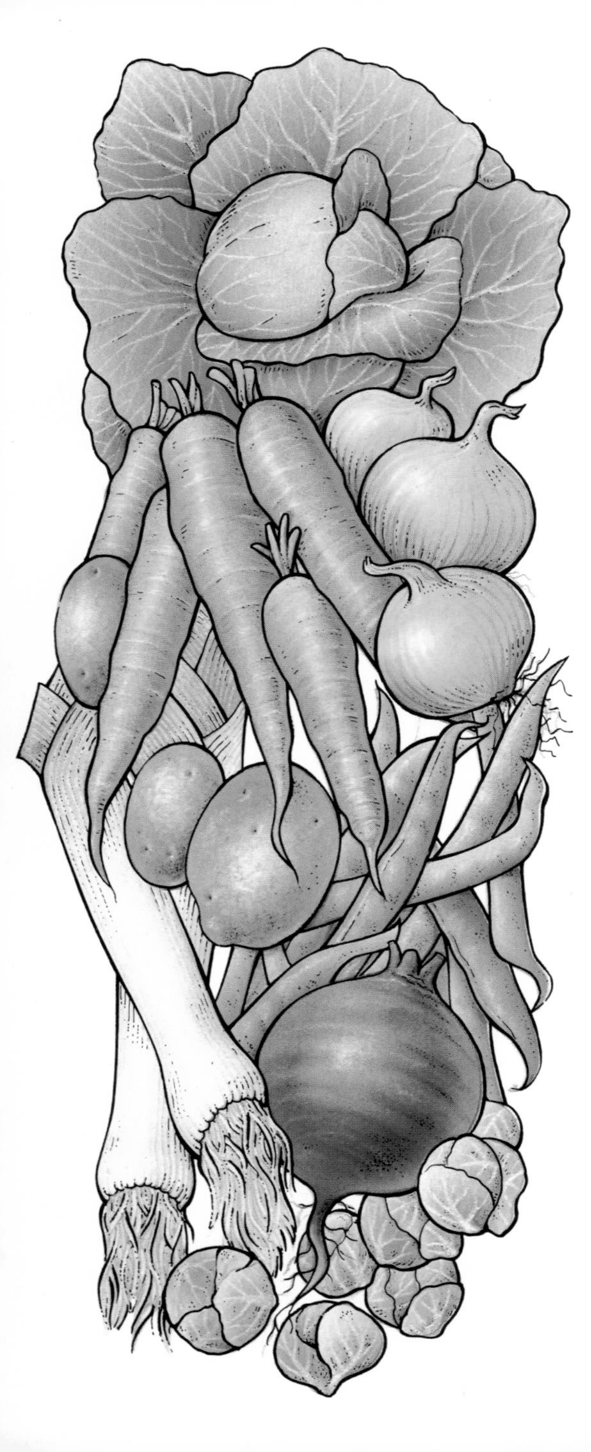

All words that appear in **bold** are explained in the glossary on page 30

First published in the U.S. in 1990 by Carolrhoda Books, Inc.
 First published 1989 by Wayland (Publishers) Ltd.

Library of Congress Cataloging-in-Publication Data

Wake, Susan.
Vegetables / Susan Wake ; illustrated by John Yates.
p. cm.—(Foods we eat)
Includes index.
Summary: Describes different types of vegetables, their history, how they are grown, and their role in human diet and health. Includes recipes.
ISBN 0-87614-390-7 (lib. bdg.)
1. Vegetables—Juvenile literature. 2. Truck farming—Juvenile literature. 3. Cookery (vegetables)—Juvenile literature. [1. Vegetables. 2. Cookery—Vegetables.] I. Yates, John, ill. II. Title. III. Series: Foods we eat (Minneapolis, Minn.)
SB324.W35 1990
635—dc20 89-9697
CIP
AC

Printed in Italy by G. Canale C.S.p.A., Turin
Bound in the United States of America

1 2 3 4 5 6 7 8 9 10 99 98 97 96 95 94 93 92 91 90

Contents

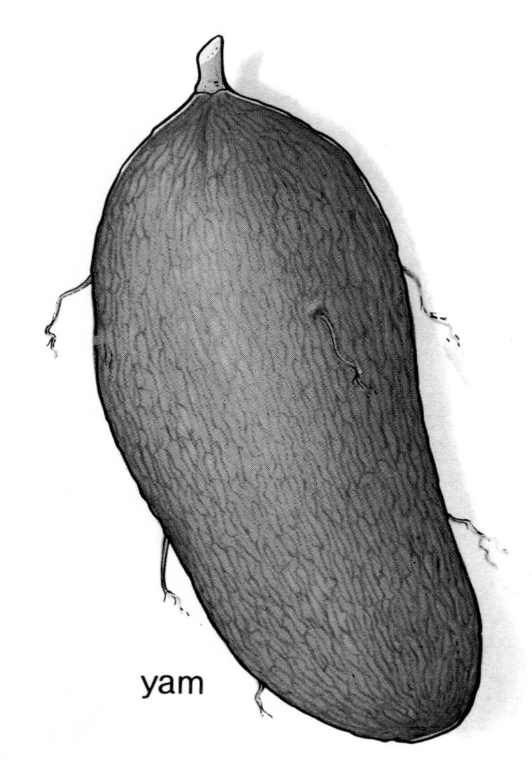

What is a vegetable?

A vegetable is the part of a plant that we eat. It may be the root, stem, leaf, seeds, or even the flowers and buds of the plant. Some vegetables are eaten raw, others are cooked, and many can be eaten either way.

Some of the vegetables we eat grow wild, but most are specially raised. Some vegetables, such as potatoes, are easy to grow and are found in

many countries. Others, such as yams, only grow in tropical climates. Certain vegetables, such as cabbages, thrive in cool conditions.

Vegetables can now be produced all over the world, even in places where they do not grow naturally. Plants that need warmth to thrive can be raised in enclosed buildings called **greenhouses.** **Irrigation** brings extra water to dry areas so crops can grow. Because of these and other farming methods, the vegetables in our stores may come from almost any part of the world.

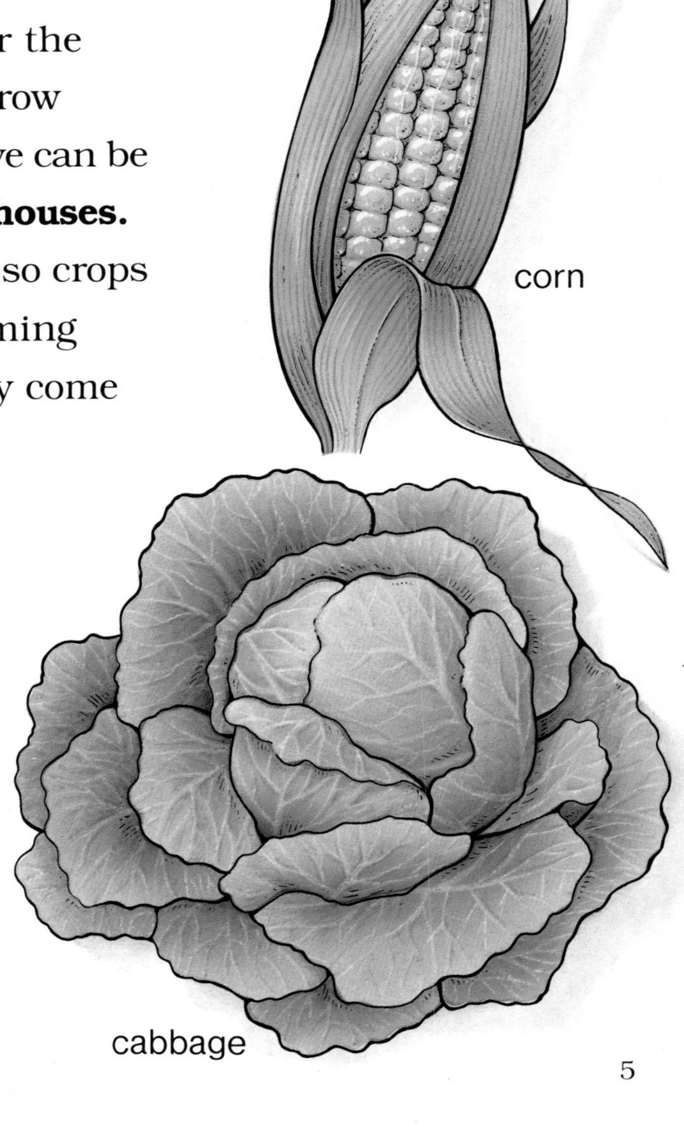

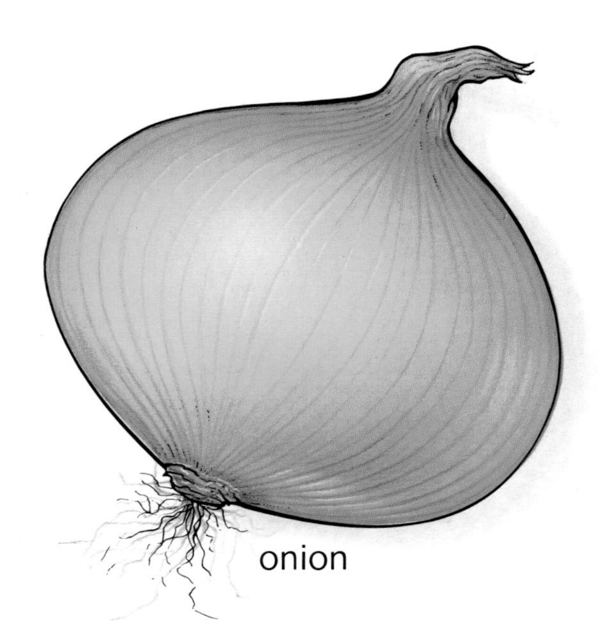

Different types of vegetables

All vegetables come from some part of a plant. **Root vegetables** come from plants that store food in their roots. The roots swell, and the swollen roots are the vegetables that we eat. Carrots, beets, and radishes are all root vegetables.

Some plants store food in special underground stems. These stems swell to form **tubers**, which we dig up and eat. Potatoes, yams, and Jerusalem artichokes are all tubers.

We may eat the leaves, stems, flowers, or buds of some plants. Asparagus and leeks are stems; Brussels sprouts, cabbages, and lettuces are leaves. Cauliflower and broccoli are clusters of flower buds.

Peas and some types of beans are seeds. When we eat green beans, we are eating both seeds and pods. Corn kernels are also seeds.

There are several different varieties of each vegetable. Next time you go shopping, look at the selection available in the grocery store.

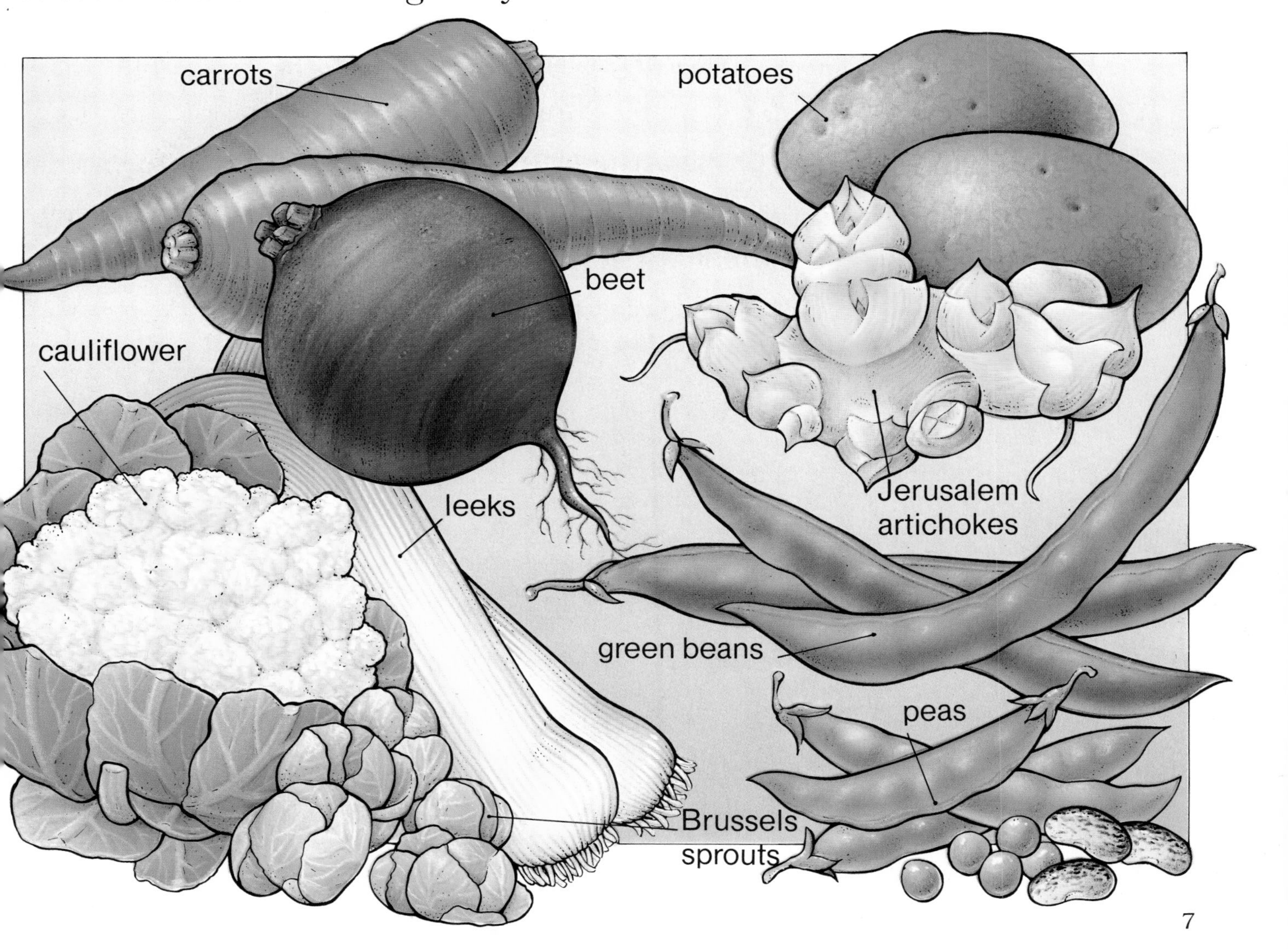

Fruits or vegetables?

Some foods that we call vegetables are in fact fruits. The fruit is the part of the plant that contains the plant's seeds. If you can see the seeds inside a "vegetable," it is really a fruit. For example, tomatoes, peppers, and cucumbers are all fruits.

Tomatoes, peppers, and cucumbers are all fruits.

Left: Packing tomatoes into boxes on a Moroccan farm. They are now ready to be shipped to markets.

Below: Choosing vegetables from a colorful selection on sale at a market in Bolivia

We call them vegetables because we tend to think of fruits as sweet. Some foods that we think of as fruits, such as watermelons, are actually considered vegetables by scientists who study plants.

Vegetables in the past

Vegetables have been an important part of people's diets for thousands of years. In early times, vegetables were not thought to be necessary for a healthy diet, but today we know that they are.

The first humans ate the fruit and roots of wild plants. Later, people collected seeds and planted them to grow food. They made primitive tools to help farm these plants. Eventually, as people

This illustration shows a medieval farmer plowing his fields in preparation for planting a vegetable crop.

migrated, they brought their plants to other parts of the world.

Onions and beans have been found in ancient Egyptian tombs. The ancient Egyptians believed that the dead would need these vegetables in their next lives.

The ancient Romans ate leeks, onions, lettuce, cauliflower, olives, celery, beans, and peas. They brought these plants to many parts of the world as they conquered more and more countries.

It was not until the beginning of the 20th century that people realized our bodies need vitamins to stay healthy. Vegetables are an excellent source of vitamins.

Vegetables for health

It is important to have a balanced diet. We need to eat a variety of foods to give us the important carbohydrates, proteins, fats, minerals, and vitamins we need.

Although vegetables contain a great deal of water, they are an excellent source of vitamins and minerals. Vegetables also provide **fiber**, which helps our digestive systems to function properly.

This diagram shows the approximate amounts of different nutrients contained in vegetables.

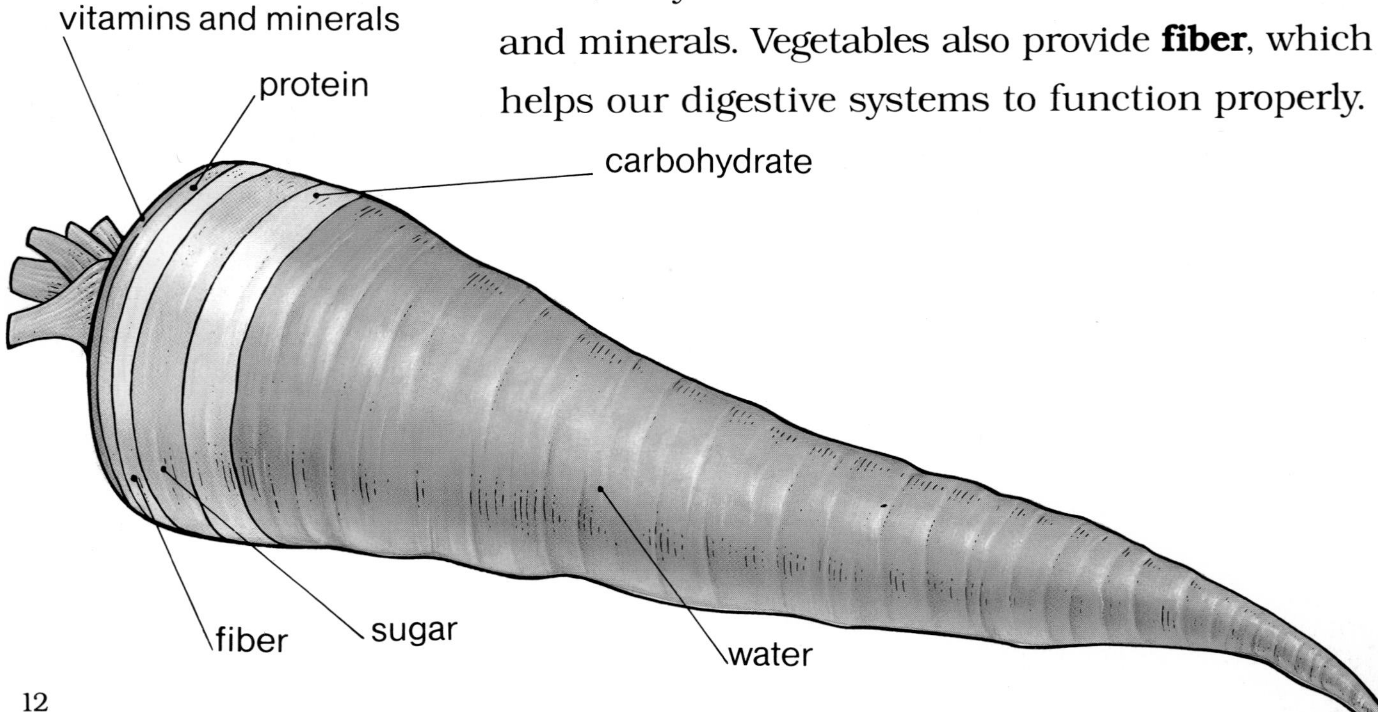

Right: Preparing a salad. Fresh raw vegetables are very nutritious.

Below: Preparing a meal at a vegetarian restaurant

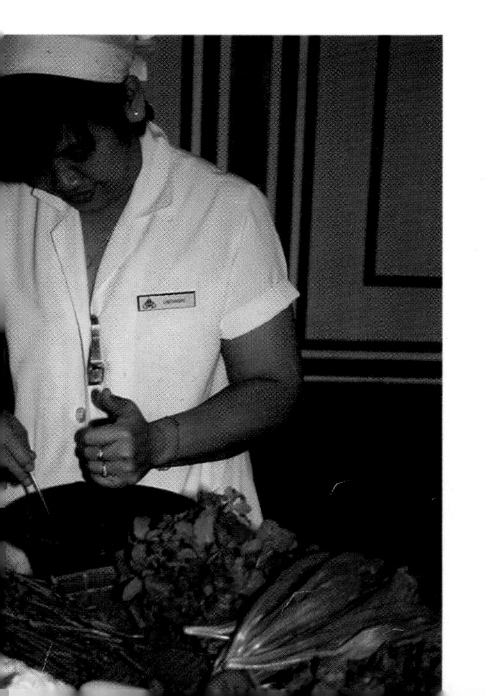

Most vegetables contain very little fat.

When vegetables are overcooked, the minerals and vitamins in them are lost. Vegetables that are raw or lightly cooked are more nutritious than those that are cooked for a long time.

Processes such as freezing, canning, and drying can also destroy some of the nutrition in vegetables.

Growing vegetables

Many people grow vegetables in home gardens, but most vegetables are grown on **truck farms**, farms which produce only vegetables.

Vegetable growers often add **fertilizers** to the soil to help plants grow. Growing plants are often sprayed with **pesticides** and **fungicides** to protect them from pests and diseases. Some farmers prefer not to use these chemicals because they

A farmer working on his vegetable plot in Nanjing, China

Left: Watering cabbages on a farm in Thailand

Below: A farmer sprays his vegetable crop with pesticides. Some farmers prefer not to use these chemicals; their produce is known as "organic."

can be harmful.

Many truck farms have greenhouses. Greenhouses are used in cool climates to protect growing plants from frost, which can kill them.

Farmers often plant seeds in batches so their vegetables will ripen at different times. This way, there are several harvests instead of just one. Once harvested, the crops are sorted, graded according to quality, and packed. This **produce** is then shipped to factories to be canned or frozen, or to markets and stores to be sold.

Vegetables and celebrations

Yam festivals are traditional in many parts of Africa and the South Pacific. The best yams are offered to the god of the harvest.

People in many parts of the world hold harvest festivals to celebrate the gathering of the crops that they have worked hard to grow. At harvest festivals in Europe, vegetables and fruits are displayed in churches, and special services are held.

Vegetables have become associated with

certain holidays and traditions. For example, on October 31, Halloween celebrations take place in some countries. On Halloween, children hollow out pumpkins to make jack-o-lanterns. In the United States, pumpkin pie has become a traditional part of Thanksgiving celebrations, and Indian corn is often used as a decoration at Thanksgiving time.

Making jack-o-lanterns out of pumpkins to celebrate Halloween in Ontario, Canada

Vegetables all year long

When vegetables are harvested, some are sold to be eaten fresh, but a large part of the crop will be eaten later in the year. Fresh vegetables will eventually begin to rot, so they must be **preserved** in order to be eaten later.

Freezing is a widely used method of preserving vegetables. Freezing must be done very quickly so

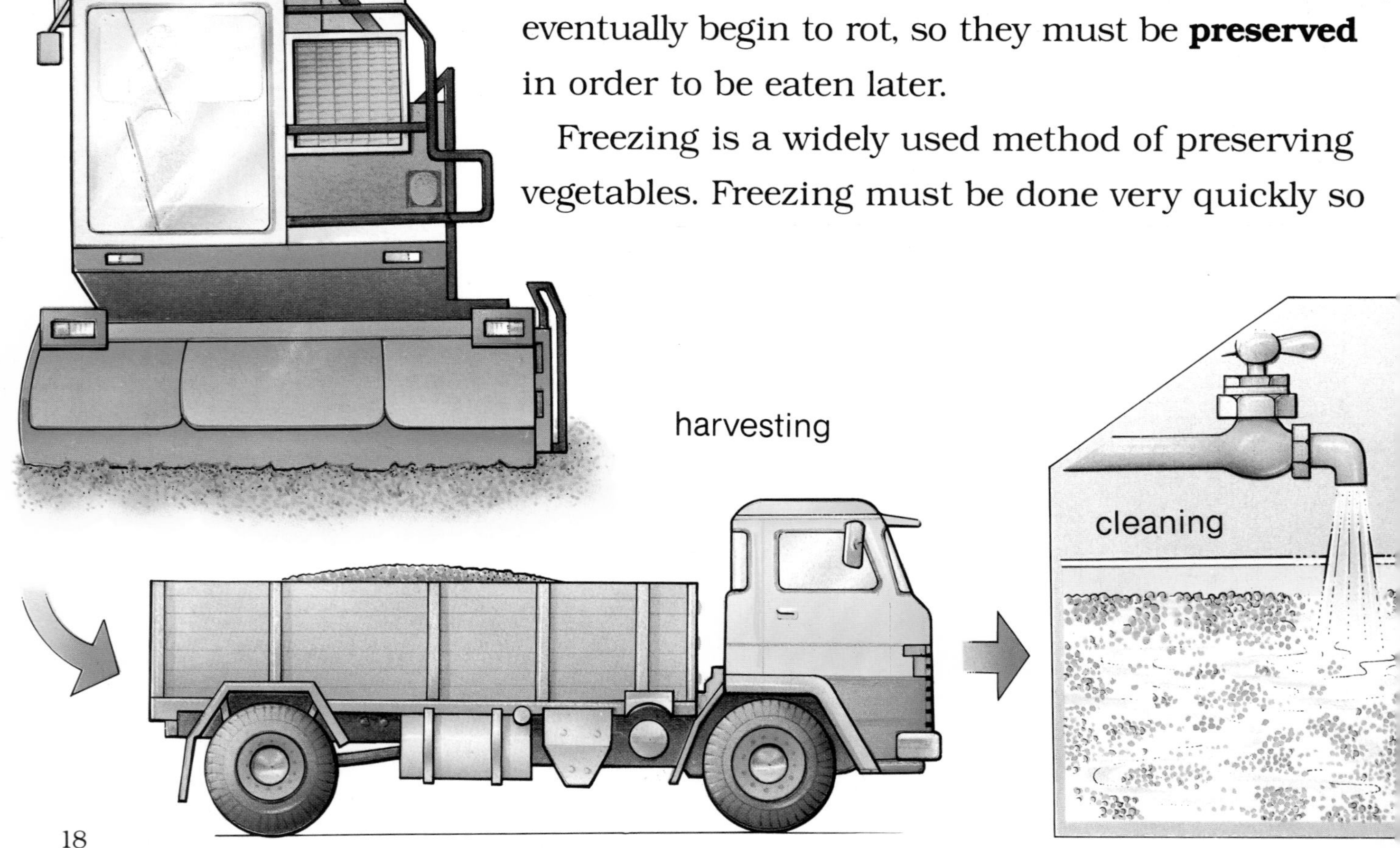

that the vegetables will remain fresh. Vegetables are cleaned and **blanched**, or boiled briefly, before they are frozen. Blanching preserves the taste of the vegetables and keeps nutrients from being lost. Frozen vegetables are the most nutritious type of preserved vegetables.

Drying is an ancient way of preserving food. It is also a good method of keeping in nutrition. Vegetables are washed and blanched before they are dried. Then they are placed in insulated

This diagram shows how frozen peas are processed.

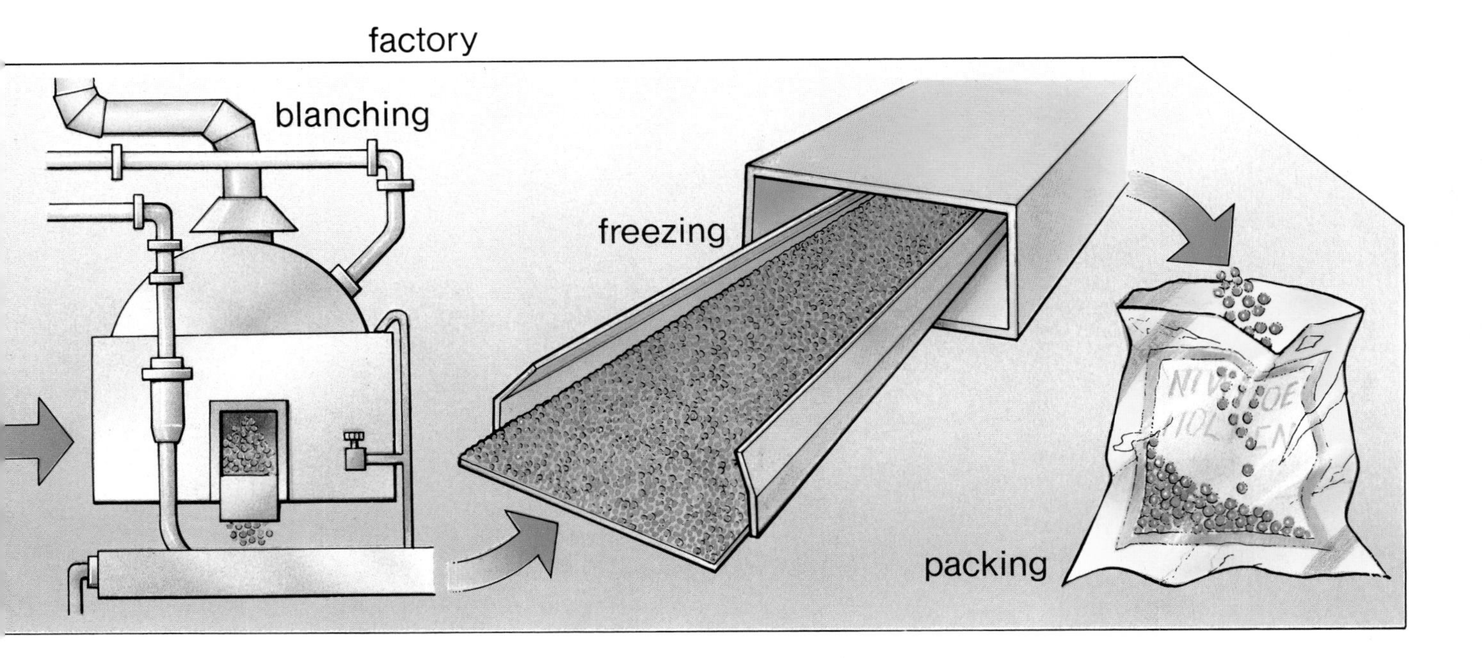

cabinets, and hot air is blown over them until they are completely dry. Dried vegetables may be used in soup mixes or packaged meals.

Canning vegetables is another way of preserving them. Some nutrition is lost in the canning process, but canned vegetables are popular and convenient.

Some people do their own canning at home, but most canning is done in factories. In canning

Above: Baked beans are among the most popular canned vegetables.

Left: Vegetables may be preserved by pickling.

Modern farming methods, such as using plastic tunnels that provide sheltered conditions, allow vegetables to be grown all year long.

factories, vegetables are sterilized so that they are perfectly clean. Machines then put them in cans. The vegetables are cooked in the cans, which are then sealed and labeled.

Vegetables may also be preserved by being pickled in vinegar. Some people do this at home with vegetables from their gardens.

These different methods of preserving food make it possible for us to eat vegetables that have been harvested many months before.

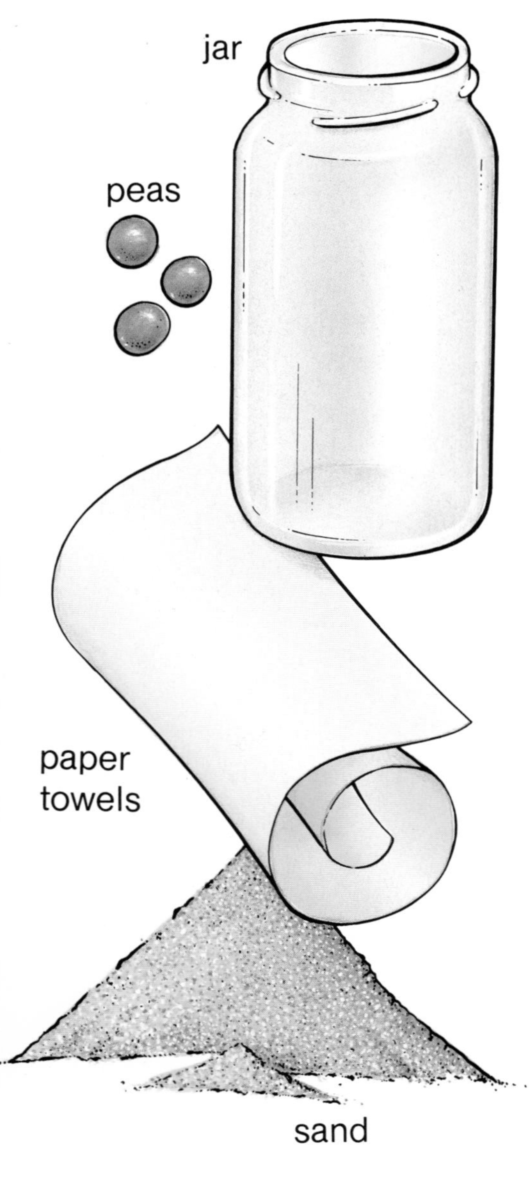

Grow your own peas

You will need: a clear jar, paper towels, sand, two or three pea seeds that have been soaked in water for 24 hours, a popsicle stick, and water.

Put the paper towels around the inside of the jar. Pack the jar with sand. Using the popsicle stick, push the soaked seeds down between the sides of the jar and the paper. Add enough water to the

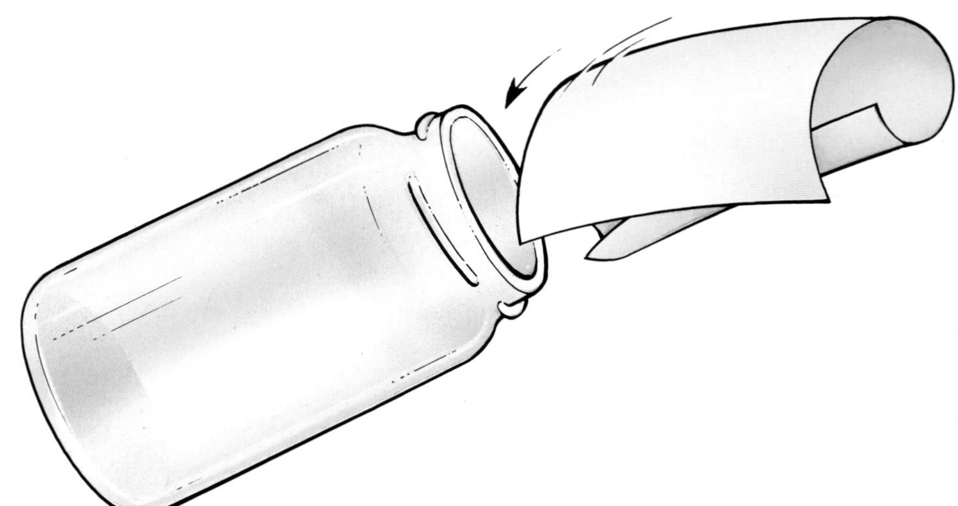

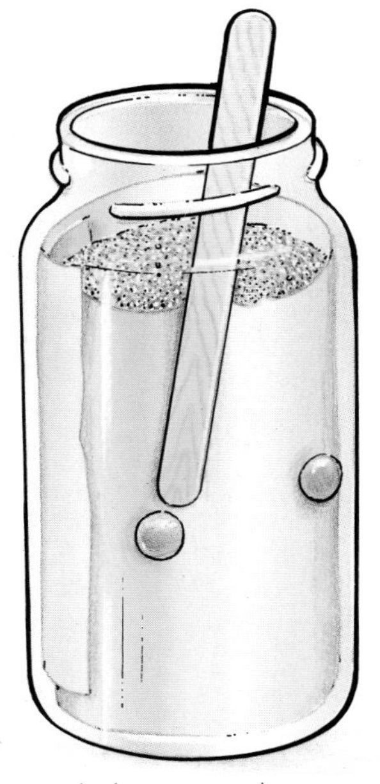

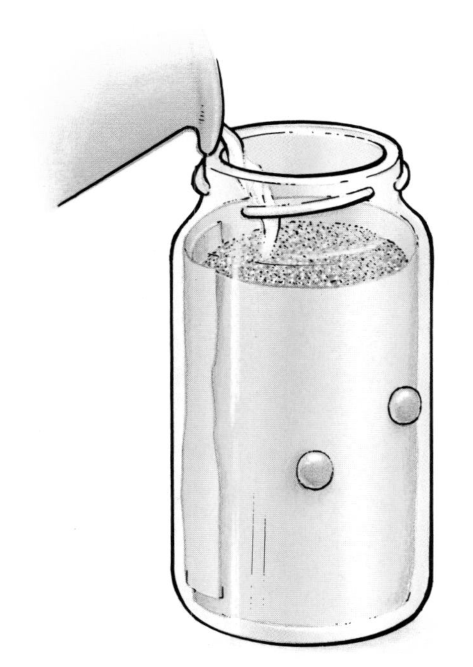

sand to keep the paper moist but not too wet. Place the jar in a warm, light place.

Observe the progress of the peas every day. Note whether or not the plumule, or young shoot, appears before the radicle, or young root. Add water to the sand as needed.

When the plant has begun to look fairly strong, try transplanting it into a small pot of soil. Be sure to keep the soil moist. Eventually, if it is the right time of year, you can plant it outdoors and produce your very own crop of peas.

Vegetables in the kitchen

There are many ways of preparing and cooking vegetables. It is important that vegetables are cooked properly so that they keep their flavor and nutritional value.

Vegetables should be peeled thinly so that the

Above: Peel vegetables thinly.

Right: A delicious dish—stir-fried vegetables

Left: In India, many people are vegetarians. This man is selling vegetables by the roadside.

Raw vegetables are the most nutritious, since they do not lose any nutrients during cooking.

nutrients concentrated just under the skin are not peeled away. You may leave the peel on some vegetables, such as carrots and cucumbers, if you scrub them well. Prepare and cook vegetables just before serving them so that they do not lose vitamin C. Cook them until they are just tender, since overcooking destroys vitamins.

Many people around the world are **vegetarians**. Some people do not eat meat for religious reasons, others because they feel it is wrong to eat animal flesh. Many cannot afford to buy meat.

Make sure there is an adult nearby when you are cooking.

Russian salad

You will need, for 3-4 people:

1 head of lettuce
1 cup cooked potatoes
1 cup cooked carrots
½ cup cooked peas
½ cup cooked green beans
mayonnaise
1 hard-boiled egg
4 dill pickle slices

1. Wash the lettuce and shake the leaves dry. Arrange them in a salad bowl.

2. Cut the potatoes and carrots into cubes and put them into a large bowl.

3. Add the peas and beans and mix well.

4. Mix gently with the mayonnaise, using enough to coat the vegetables.

5. Pile the mixture on top of the lettuce. Garnish with slices of hard-boiled egg and pickle.

Corn fritters

You will need, for 3-4 people:

½ cup whole wheat flour
1 egg, beaten
½ cup milk
1 cup fresh or frozen corn
2 tablespoons oil

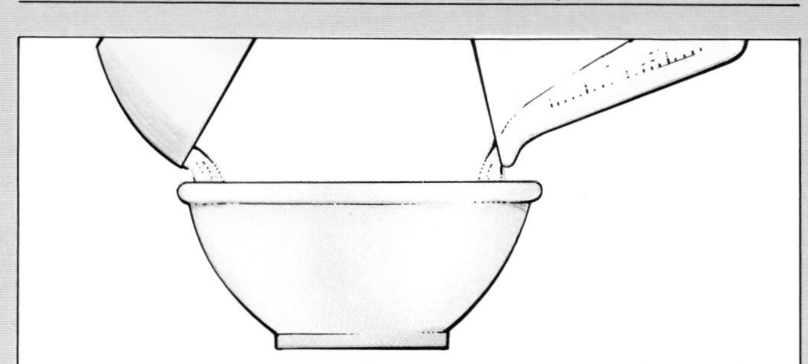

1. Put the flour in a bowl. Add the egg and half the milk.

2. Beat until smooth. Gradually add the rest of the milk and beat well to make a batter.

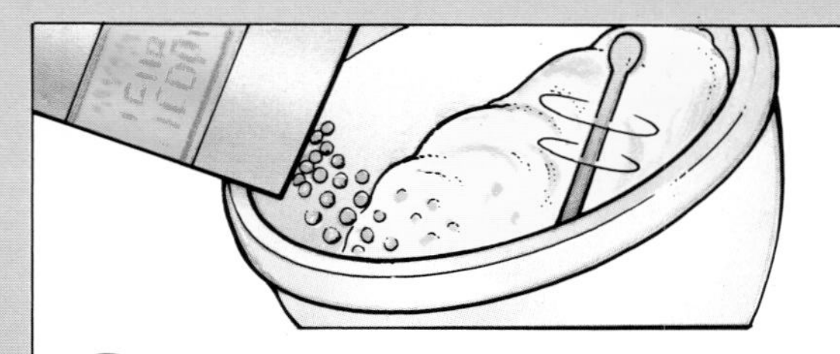

3. Stir in the corn.

4. Heat the oil in a frying pan, and carefully drop the batter by spoonfuls into the hot oil.

5. Fry for 3-4 minutes on each side, until the fritters are golden brown.

Vegetable broth

You will need:

1 carrot
1 parsnip
half a turnip
2 celery stalks
1 leek
1 onion
1 tablespoon butter
3 cups water
1 tablespoon barley (rinsed)
1 teaspoon salt

1. Peel the carrot, parsnip, and turnip. Wash them well, then cut them into small cubes.

2. Wash the celery and leek, then chop them. Peel and dice the onion.

3. Melt the butter in a saucepan. Add the vegetables and cover the saucepan.

4. Fry the vegetables without browning them for about 7 minutes, shaking the pan.

5. Pour the water into the saucepan. Add the barley and salt and bring to a boil.

6. Lower the heat and put the lid on. Simmer for about 45 minutes or until the barley is soft.

Glossary

blanched: scalded or boiled quickly, but not completely cooked

fertilizers: natural or artificial materials added to soil to help plants grow

fiber: the coarse, bulky material in some foods that helps keep our digestive systems healthy

fungicides: chemicals that are used on plants to keep them from getting diseases

greenhouses: buildings made of glass in which plants are grown. In cold climates, plants can be grown in greenhouses to be protected from frost.

irrigation: the process of bringing water to dry areas

pesticides: chemicals that are used on plants to kill insects

preserved: kept fresh through various methods, such as canning and freezing

produce: fresh fruits and vegetables

root vegetables: vegetables, such as carrots, that are the roots of certain plants

truck farms: farms on which only vegetables are grown

tubers: the edible underground stems of certain plants. Vegetables such as potatoes are tubers.

vegetarians: people who eat no meat

Index

Photo acknowledgments

The photographs in this book were provided by: pp. 9 (left), 13 (bottom), 20 (left), Christine Osborne; p. 9 (right), Hutchison Library; p. 10, Wayland Picture Library; pp. 11, 13 (top), 15 (right), 17, 21, Zefa; p. 15 (left), Holt Studios; p. 20 (right), H.J. Heinz Company; pp. 24, 25, J. Allan Cash.